YOUR KNOWLEDGE HAS VALUE

- We will publish your bachelor's and master's thesis, essays and papers

- Your own eBook and book - sold worldwide in all relevant shops

- Earn money with each sale

Upload your text at www.GRIN.com
and publish for free

Bibliographic information published by the German National Library:

The German National Library lists this publication in the National Bibliography;
detailed bibliographic data are available on the Internet at http://dnb.dnb.de .

This book is copyright material and must not be copied, reproduced, transferred,
distributed, leased, licensed or publicly performed or used in any way except as
specifically permitted in writing by the publishers, as allowed under the terms and
conditions under which it was purchased or as strictly permitted by applicable
copyright law. Any unauthorized distribution or use of this text may be a direct
infringement of the author s and publisher s rights and those responsible may be
liable in law accordingly.

Imprint:

Copyright © 2016 GRIN Verlag, Open Publishing GmbH
Print and binding: Books on Demand GmbH, Norderstedt Germany
ISBN: 9783668353619

This book at GRIN:

http://www.grin.com/en/e-book/344988/euler-s-number-why-is-eule-s-number-e-the-basis-of-natural-logarithm

Sumaanyu Maheshwari

Euler's number. Why is Eule's number "e" the basis of natural logarithm functions

GRIN Publishing

GRIN - Your knowledge has value

Since its foundation in 1998, GRIN has specialized in publishing academic texts by students, college teachers and other academics as e-book and printed book. The website www.grin.com is an ideal platform for presenting term papers, final papers, scientific essays, dissertations and specialist books.

Visit us on the internet:

http://www.grin.com/

http://www.facebook.com/grincom

http://www.twitter.com/grin_com

Why is the Euler's number 'e' is the base of natural logarithm functions?

Sumaanyu Maheshwari

Introduction

When the concept of logarithms was first introduced to me, a plethora of questions revolved around my mind. My inquisitiveness compelled me to think and ask questions as to where are the practical applications of logarithms, why do we take different bases of these functions and what is the need for natural logarithms. Amongst these questions, one particularly intrigued me: why is e particularly the base of the natural logarithm. Why out of all numbers that exist did we choose e as the base of the natural logarithm function? I was fascinated by why taking the base e made the normal logarithm a natural logarithm.

Therefore, to quench the curiosity of many others like me, I will show through this paper that why e is the correct choice for the base of exponential and natural logarithm functions. I shall also be exploring the most important property of e, via this paper.

Theory

About e

The constant e is a real and an irrational number that has a value, which is approximately equal to 2.71828, when given up to 6 significant digits. Like π, as proved by Charles Hermite, e is a transcendental number[1]. A transcendental number is either a non-algebraic complex or real number, which is not a root of any non-zero rational polynomial equation. It is most commonly seen as the base of natural exponential function and at the base of a natural logarithm function. $f(x) = e^x$

Exponential and Logarithm functions

A natural exponential function is a certain kind of function where e is multiplied x times with itself, which intern can be written as x raised to the power e. Therefore, $f(x) = e^x$, is a natural exponential function. This process of raising powers is called exponentiation. Logarithms are the inverse operation to this process of exponentiation. This means that the logarithm function of a number is a certain kind of function in which the exponent is raised to a certain base as to produce the required number. For example when the logarithm of x to the base b gives y ($log_b x = y$), then, $b^y = x$ where $b > 0 \ and \ b \neq 1$.

---

[1] Pickover, Cliff. "The 15 Most Famous Transcendental Numbers – Cliff Pickover". *Sprott.physics.wisc.edu.* Web. 2015.

When the base of the logarithm function is e then it is termed as a natural logarithm function. When the base of the logarithm function is 10 then it is termed as common logarithm. As astonishing as it sounds these logarithms with different bases have different uses. The common logarithm is most efficaciously and commonly used in spectroscopy and in various engineering fields, whereas the natural logarithm function is generally used in statistics and economics.

The rationale behind the use of natural logarithm instead of common logarithm is justified in the Andrew Gelman and Jennifer Hill's book on regression. Under the section of the concept of linear regression of social sciences, it was stated that, "We prefer natural logs (that is, logarithms base e) because, as described above, coefficients on the natural-log scale are directly interpretable as approximate proportional differences: with a coefficient of 0.06, a difference of 1 in x corresponds to an approximate 6% difference in y, and so forth."[2]

This means that $\exp(x) \approx 1 + x$[3], in Taylor series expansion for very small values of x. So, in the example given in the book there is difference of 0.06 on natural logarithm scale, which corresponds to an approximate multiplication of 1.06 on their original scale. This gives a 6% increase.

However, the Taylor series expansion of common logarithms or rather exponents of 10 is comparatively convoluted, which is given by:

$$10^x \approx 1 + 2.302585x$$

In this case 2.302585 is the value of the natural logarithm of 10. This goes on to state that a change of 0.01 on a common logarithm scale will approximately correspond to increment of 2.3% on the original scale.

I shall be giving more details about the applications of natural logarithm functions and of e later in this paper.

History of e:

The first reference to e was in 1618 by John Napier in a table of an appendix for his work on

---

[2] Gelman, Andrew, and Jennifer Hill. *Data Analysis Using Regression and Multilevel/Hierarchical Models*. 1st ed. New York: Cambridge UP, 2007. 60-61. Print.
[3] Cook, John D. "Another Reason Natural Logarithms Are Natural." *Data Analysis*. John D. Cook, 5 Feb. 2015. Web. 2015. <http://www.johndcook.com/blog/2015/02/05/natural-logarithms-are-natural/>.

logarithms. William Oughtred, as many believe, wrote this table. In 1647 the area under a rectangular hyperbola was a found out by Saint Vincent[4]. However many believe that he was unaware of the connection of this with logarithms. It was in 1661 that Christiaan Huygens found out the relationship between logarithms and the rectangular hyperbola ($y \times x = 1$)[5] He also defined a new curve, which he called as logarithmic, but it actually was an exponential curve, as we know today. He also calculated the value of log of e as taken to the base 10, up to 17 decimal places, but it appeared to be a calculation of a certain constant, other than e. So up to 1661, although the number e was used allusively and seen in certain papers and theories it was not explicitly defined or calculated. Following in 1668, there was development on logarithms where Nicolaus Mercator calculated the expansion of *log(1+x)* in his Logarithmotechnia. He described log to the base e as natural logarithms. But there was no explicit appearance of e yet again. It was only in 1683 that Jacob Bernoulli through his study on compound interest, and not logarithms, discovered the number e. It is said that he tried to solve a problem in compound interest by calculating a limit of $(1 + \frac{1}{n})^n$, where n tends to infinity and the limit existed between 2 and 3, via binomial theorem. However, he failed to see any connections between logarithm and his results. In 1684 James Gregory made connections between exponents and logarithms. In 1690, e explicitly appeared as the alphabet *b*. "e" first made its appearance as "e" only in 1731 in a letter when Leonhard Euler wrote to Christian Goldbach. In 1748 in his publication *'Introductio in Analysin infinitorum'*, he showed his various ideas regarding e. He presented an idea that

$$e = 1 + {}^1/_{1!} + {}^1/_{2!} + {}^1/_{3!} + \ldots$$

Where $e = \lim_{n \to \infty} \left(1 + \frac{1}{n}\right)^n$, $n \to \infty$. And calculated the value of *e* till 18 decimal places. Some properties of "*e*" that make it important and serve as a reason as to why it is considered natural and a base for the natural logarithm function. It is now known to us that e is such a number that makes the area under the rectangular hyperbola from 1 to e equal to 1. It is this property of e that makes it the base of natural logarithm function[6]. There are various other ways that e is similarly proved to be natural and thus an apt base for the natural logarithm (ln) function. I am in this paper shall be exploring another way of doing so. Before I do that I

---

[4] O' Conner, J J, and E F Robertson. "The Number E." MacTutor History of Mathematics, Sept. 2001. Web. 2015.
[5] O' Conner, J J, and E F Robertson. "The Number E." MacTutor History of Mathematics, Sept. 2001. Web. 2015.
[6] O' Conner, J J, and E F Robertson. "The Number E." MacTutor History of Mathematics, Sept. 2001. Web. 2015.

shall discuss some properties of e that make it important and makes it a reason as to why it is considered natural and a base for the natural logarithm function.

The derivative of e^x is e^x itself. This is a unique property of e and no other function can have its derivative as the function itself. Also thought the equation $e^{(i \times \pi)} + 1 = 0$, 5 most important number in mathematics are linked and contains fundamental concepts like addition, multiplication, raising to power and equality. 'e' is also liked with calculus; through Euler's equation, which he deduced by the de moiveres formula, $e^{ix} = \cos(x) + i\sin(x)$. This equation also leads to Fourier analysis. There are so many other important properties, but the last property I shall be giving is:

$$e = \lim_{n \to \infty} \left(1 + \frac{1}{n}\right)^n$$

In recognition of the natural description of the properties of 'e', the exponential functions and logarithmic functions are called natural.

Overview

For an exponential curve, since there is always going to be a tangent, which is dependent on the base of the function, at every point on the curve, I shall be finding out the base of the exponential function for which the tangent can be equal to 1. By doing so I will obtain a function whose derivative will be equal to the function itself. Such a function can be none other than 'a^x'. However, so as to mathematically find such a base, I shall be considering a function a^x. I will take close positive and negative estimates for this base 'a', and ultimately prove that the true value of 'a' for the desired conditions will be given by the common limit of the monotone increasing and monotone decreasing functions. This natural base of exponential functions is also used as the base for the logarithm functions, thus naming it as the natural logarithm function.

Proof:

I hope to find the value of e using the slope of exponential functions. It is evident from all the exponential functions that for all bases these functions are convex, this means that if we join any two points on the curve, the line segment joining two such points will always be above the curve. Formally expressing this property of exponential curves:

Let there be a point A with coordinates (x_1, a^{x_1}) and a point B with coordinates x_2, a^{x_2} for some exponential curve $y = a^x$. Now, suppose that there is a point C at lies in the interior of the curve in such a way that it lies on the line segment $\overline{AB}$ that divides $\overline{AB}$ in ratio $p:q$, where p and q are positive numbers and $(p + q = 1)$ in such a case the coordinates of C will be $(px_1 + qx_2, a^{x_1} + a^{x_2}.)$ I took $p + q = 1$ as, when the section formula is applied the denominator would have $p + q$, so to simplify the calculations while finding the coordinates of C, $p + q$ is taken as 1. For these coordinates of C there will always lie a point D on the curve, directly below C, that will have coordinates $(px_1 + qx_2, a^{px_1+qx_2})$. Therefore, for convexity, I can correctly say that $a^{px_1+qx_2} \leq a^{x_1} + a^{x_2}$.

Graph 1 depicting the convexity

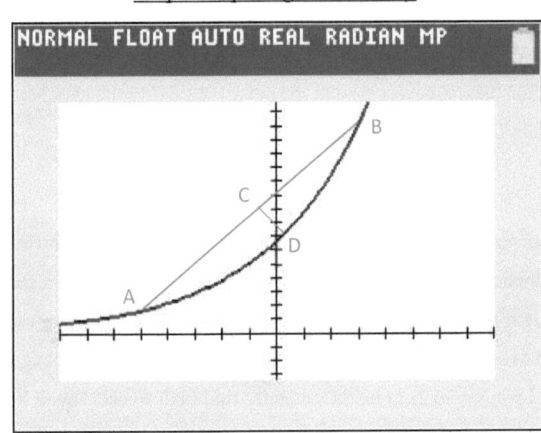

The slope of the exponential function has no breaks and is always continuous. I can say this because the differentiation of the function gives the slope[7] and therefore $\frac{d}{dx}a^x = a^x \ln a$[8] Now it is evident that both the exponential function a^x are continuous and have no breaks and $\ln a$ is a constant; thus they are defined at all points. Using this information, I can deduce

---

[7] Zidar, Owen. "A Primer on Derivatives and Maximization Problems." *A Primer on Derivatives and Maximization Problems 1* (n.d.): n. pag.*Faculty.chicagobooth*. Chicagobooth, 2015. Web. 28 2015.
<http://faculty.chicagobooth.edu/owen.zidar/teaching/Fall%202015/Week0/Week0_Primer_On_Derivatives_and_Maximization.pdf>.
[8] Kahen. "Show That D/dx (a^x) = A^xln A." *Calculus*. Stack Exchange Inc, May 2013. Web. 2015. <http://math.stackexchange.com/questions/398826/show-that-d-dx-ax-ax-ln-a>.

that there will be a tangent at every point of the curve. The slope of the tangent depends on the base of the function.[9]

To prove this, Let there be a function $f(x) = a^x$ and m_a be the gradiesnt of the graph at $x = 0$. This means that the derivative of the function at point $x = 0$ is m_a. This means that $f'(x) = m_a$.

Now, using the first principle, $f'(x) = \lim_{h \to 0} \frac{f(x+h)-f(x)}{h}$ [10]

$$= \lim_{h \to 0} \frac{a^{(x+h)} - a^x}{h}$$

$$= \lim_{h \to 0} \frac{a^x a^h - a^x}{h}$$

$$= \lim_{h \to 0} a^x \frac{a^h - 1}{h}$$

$$= \lim_{h \to 0} a^x \frac{a^h - a^0}{h}$$

$$= \lim_{h \to 0} a^x \frac{f(0+h) - f(0)}{h}$$

$$= a^x \lim_{h \to 0} \frac{f(0+h) - f(0)}{h}$$

$$= a^x f'(0)$$

$$= a^x m_a$$

Therefore, I can write that

$$f'(x) = f(x) m_a$$

With this established, I am going to find that what should be the base of the exponential function to make the slope of the tangent at the coordinate (0,1) equal to 1. I am going to take

---

[9] "Derivative of Exponential Functions." *Derivative of Exponential Functions*. Geogebra, Web. 2015. <http://webspace.ship.edu/msrenault/GeoGebraCalculus/derivative_exponential_functions.html>.

[10] Martin, David. "Differentiation from First Principles." *Mathematics for the International Student: Mathematics HL (Core)*. 3rd ed. Adelaide: Haese Mathematics, 2012. 523. Print.

the coordinates (0,1) because for any exponential function with a general formula $y = a^x$, will pass through (0,1) for all value of a.

This means that I am going to find such a base of the exponential function that will just touch the line $y = x + 1$. This is so because I am going to find out an exponential function whose tangent, which is the gradient of the function, is 1. And this straight-line equation has a gradient of 1, so they will just touch.

In more mathematical terms this will mean that for this base, the derivative of the base raised to the power x will be the same as the base raised to the power x. Let us suppose this base is 'a'.

Further, I justify taking the value of gradient as 1 at (0,1) specifically due to the following reason:

Suppose the function is given by $f(x) = a^x$

Now using the first principle, the derivative will be:

$$f'(x) = \lim_{h \to 0} \frac{a^{(x+h)} - a^x}{h}$$

$$= \lim_{h \to 0} \frac{a^x (a^h - 1)}{h}$$

Now putting the value of $x = 0$ in $f(x)$

$$f'(0) = \lim_{h \to 0} \frac{f(0 + h) - f(0)}{h}$$

$$= \lim_{h \to 0} \frac{a^x - 1}{h}$$

From these equations I can say that

$$f'(x) = a^x f'(0)$$

And

$$f(x) = a^x$$

Since I want to prove that the base I choose will have its derivative same as the base raised to the power x, therefore $f'(0)$, which is the gradient will have the value of 1. Referring to the previous equations I can say that $\lim_{h \to 0} \frac{a^x - 1}{h} = 1$.

This means I will take such a case where $f(x) = f'(x) = a^x$

To get an accurate estimate for such a value of a, I will choose points of the tangent that are very close to the decided coordinate, that is (0,1).

Estimate 1:

Suppose that x is a very large number and $x > 0$. Thus, taking a point on the right side of 0 and very close to it, x must have a value close to (0,1). This means it must be equal to $(\frac{1}{x}, 1 + \frac{1}{x})$. Since our curve $y = a^x$ follows the properties of convexity, the tangent will below the curve except at the point where it touches the curve (see graph 2 below)

From the graph and the properties of convexity $a^{\frac{1}{x}} > 1 + \frac{1}{x}$

Raising both sides with $[x]$

$$a > \left(1 + \frac{1}{x}\right)^x$$

Let's term it as equation 1

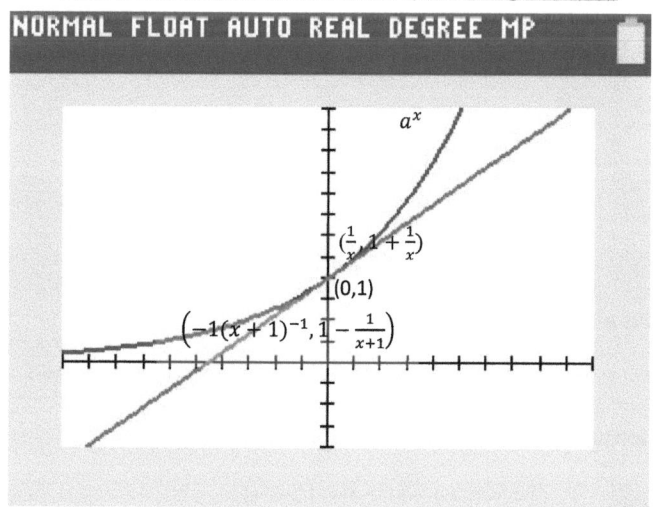

Graph 2 depicting an exponential curve and its tangent at (0,1)

I took the curve in graph 2 as 3^x. The result of this graph is acceptable for the base a, because this graph is formed without the loss of generality. The conditions for all the exponential functions are fulfilled as 3 is greater than 0 and not equal to 0.

Estimate 2:

Since x can have both positive values, let us take a point that is on the left side of 0, which means it should be negative. Also I would prefer a numerically different value in this estimate that is closer to (0,1). Therefore such a point must be $\left(\frac{-1}{x+1}, 1 - \frac{1}{x+1}\right)$.

Again this curve follows the properties of convexity (see graph 2 above), therefore:

$$a^{\left(\frac{-1}{x+1}\right)} > 1 - \frac{1}{x+1}$$

Raising both sides by $[-(x+1)]$

$$a < \left(1 - \frac{1}{x+1}\right)^{-(x+1)}$$

Now,

$$\left(1 - \frac{1}{x+1}\right)^{-(x+1)} = \left(\frac{(x+1-1)}{(x+1)}\right)^{-(x+1)}$$

$$= \left(\frac{x}{x+1}\right)^{-(x+1)}$$

$$= \left(\frac{x+1}{x}\right)^{(x+1)} = \left(1 + \frac{1}{x}\right)^{(x+1)}$$

Therefore,

$$a < \left(1 + \frac{1}{x}\right)^{x+1}$$

Let's term this as equation 2

From the equations 1 and 2, for the base a for all $x > 0$:

$$\left(1 + \frac{1}{x}\right)^x < a < \left(1 + \frac{1}{x}\right)^{x+1}$$

Let $f(x) = \left(1 + \frac{1}{x}\right)^x$ for all $x > 0$

$$f(x) = e^{\ln\left(1+\frac{1}{x}\right)^x}$$

$$f(x) = e^{x\ln\left(1+\frac{1}{x}\right)}$$

Taking it derivative

$$f'(x) = e^{x\ln\left(1+\frac{1}{x}\right)} \times \left[1 \times \ln\left(1+\frac{1}{x}\right) + x \times \frac{1}{1+\frac{1}{x}} \times (-x^{-2})\right]$$

$$= e^{x\ln\left(1+\frac{1}{x}\right)} \times \left[\ln\left(1+\frac{1}{x}\right) + \frac{(-x)^{-1}}{1+\frac{1}{x}}\right]$$

$$= f(x) \times \left[\ln\left(1+\frac{1}{x}\right) - \frac{1}{x+1}\right]$$

Now let $\left[\ln\left(1+\frac{1}{x}\right) - \frac{1}{x+1}\right] = g(x)$

Thus $f'(x) = f(x) \times g(x)$

Now differentiating $g(x)$:

$$g'(x) = \frac{1}{1+\frac{1}{x}} \times -x^{-2} - (-(x+1)^{-2})$$

$$= \frac{-x^{-2}}{1+\frac{1}{x}} + \frac{1}{(x+1)^2}$$

Now, $\frac{-x^{-2}}{1+\frac{1}{x}} = \frac{-x^{-2}}{1+\frac{1}{x}} \times \frac{x^2}{x^2} = \frac{-1}{x^2+x}$

Therefore, $g'(x) = \frac{-1}{x^2+x} + \frac{1}{(x+1)^2}$

$$= \frac{-1}{x(x+1)} + \frac{1}{(x+1)^2}$$

$$= \frac{-(x+1)+x}{x(x+1)^2}$$

$$\therefore g'(x) = \frac{-1}{x(x+1)^2}$$

11

Therefore, $g'(x) < 0$ for all $x > 0$, this means that $g'(x)$ is a decreasing function.

But $\lim_{x \to \infty} g(x) = \ln(1) - 0 = 0$

Therefore $g(x)$ is decreasing and approaches to zero from above.

This means that $g(x) > 0$

Now that $f(x) > 0$ and $g(x) > 0$, therefore $f'(x) > 0$

For a function to be a monotone increasing function, the derivative of the function must be greater to or equal to 0. $f'(x) \geq 0$ for all $x \geq 1$.[11]

Since $f'(x) > 0$, the function $f(x) = \left(1 + \frac{1}{x}\right)^x$ is monotone increasing for all values of $x > 0$.

Graph 3 of $f(x) = \left(1 + \frac{1}{x}\right)^x$

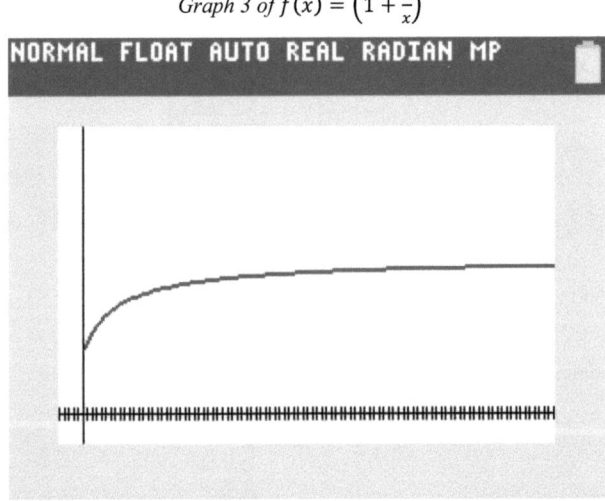

Similarly,

Let $h(x) = \left(1 + \frac{1}{x}\right)^{(x+1)}$ for all $x > 0$

$$h(x) = e^{\ln\left(1 + \frac{1}{x}\right)^{(x+1)}}$$

$$h(x) = e^{(x+1)\ln\left(1 + \frac{1}{x}\right)}$$

Taking it derivative

---

[11] Quinn, Catherine, C. J. Sangwin, R. C. Haese, and Michael Haese. "Monotone Sequences." *Mathematics for the International Student: Mathematics HL (option): Calculus, HL Topic 9, FM Topic 5, for Use with IB Diploma Programme.* N.p.: Haese, 2013. 64. Print.

$$h'(x) = e^{(x+1)\ln(1+\frac{1}{x})} \times \left[1 \times \ln\left(1 + \frac{1}{x}\right) + (x+1) \times \frac{1}{1+\frac{1}{x}} \times -x^{-2}\right]$$

Now, $(x+1)(\frac{-x^{-2}}{1+\frac{1}{x}}) = (x+1) \times \frac{-x^{-2}xx}{x+1} = -x^{-1}$

$$= e^{x \ln(1+\frac{1}{x})} \times [\ln\left(1+\frac{1}{x}\right) - x^{-1}]$$

$$= h(x) \times [\ln\left(1+\frac{1}{x}\right) - \frac{1}{x}]$$

Now, let $\left[\ln\left(1+\frac{1}{x}\right) - \frac{1}{x}\right] = k(x)$

Thus, $h'(x) = h(x) \times k(x)$

Now, differentiating $k(x)$

$$k'(x) = \frac{1}{1+\frac{1}{x}} \times -x^{-2} - (-x^{-2})$$

$$= \frac{-x^{-2}}{1+\frac{1}{x}} + \frac{1}{x^2}$$

Now, $\frac{-x^{-2}}{1+\frac{1}{x}} = \frac{-x^{-2}}{1+\frac{1}{x}} \times \frac{x^3}{x^3} = \frac{-x}{x^3+x^2}$

Therefore $k'(x) = \frac{-x}{x^3+x^2} + \frac{1}{x^2}$

$= \frac{-1}{x^2(x+1)} + \frac{1}{x^2}$

$$= \frac{-x+x+1}{x^2(x+1)}$$

$$\therefore k'(x) = \frac{1}{x^2(x+1)}$$

Therefore, $k'(x) > 0$ for all $x > 0$, this means that $k'(x)$ is an increasing function.
But $\lim_{x \to \infty} k(x) = \ln(1) - 0 = 0$

Therefore, $k(x)$ is an increasing and approaches to zero from below.
This means that $k(x) < 0$
Now that $h(x) > 0$ and $k(x) < 0$, therefore $h'(x) < 0$.

For a function to be a monotone decreasing function, the derivative of the function must be less than or equal to 0. $f'(x) \leq 0$ for all $x \geq 1$.[12]

Since $h'(x) < 0$, therefore the function $f(x) = \left(1 + \frac{1}{x}\right)^{x+1}$ is monotone decreasing for all values of $x > 0$.

$$\text{Graph 4 of } f(x) = \left(1 + \frac{1}{x}\right)^{x+1}$$

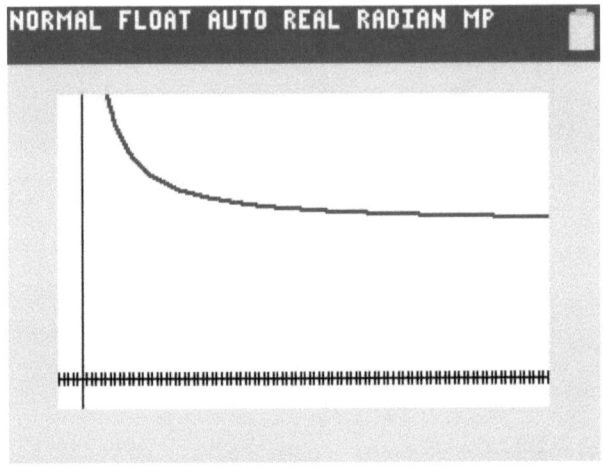

Since I want a base 'a' such that $\left(1 + \frac{1}{x}\right)^{x} < a < \left(1 + \frac{1}{x}\right)^{x+1}$ therefore I will plot the graph of $\left(1 + \frac{1}{x}\right)^{x}$ and $\left(1 + \frac{1}{x}\right)^{x+1}$ on one window.

---

[12] Quinn, Catherine, C. J. Sangwin, R. C. Haese, and Michael Haese. "Monotone Sequences." *Mathematics for the International Student: Mathematics HL (option): Calculus, HL Topic 9, FM Topic 5, for Use with IB Diploma Programme*. N.p.: Haese, 2013. 64. Print.

Graph 5 depicting $y = \left(1 + \frac{1}{x}\right)^x$ and $y = \left(1 + \frac{1}{x}\right)^{x+1}$ for a domain of x=0 to x=10

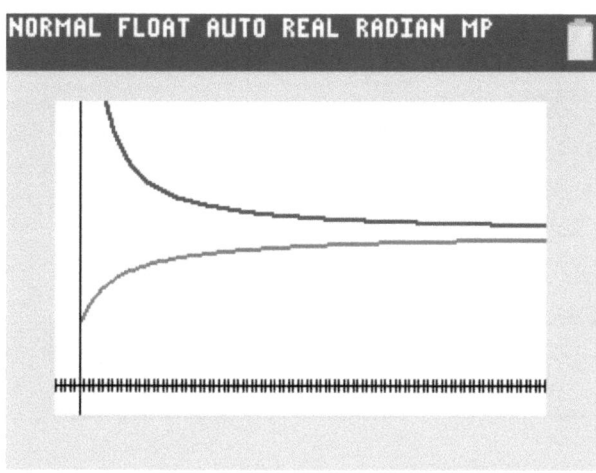

Graph 6 depicting $y = \left(1 + \frac{1}{x}\right)^x$ and $y = \left(1 + \frac{1}{x}\right)^{x+1}$ for a domain of x=0 to x=50

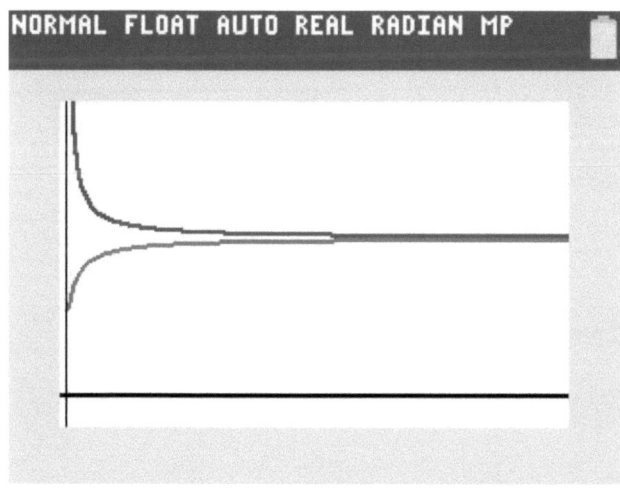

Graph 7 depicting $y = \left(1 + \frac{1}{x}\right)^x$ and $y = \left(1 + \frac{1}{x}\right)^{x+1}$ for a domain of x=0 to x=100

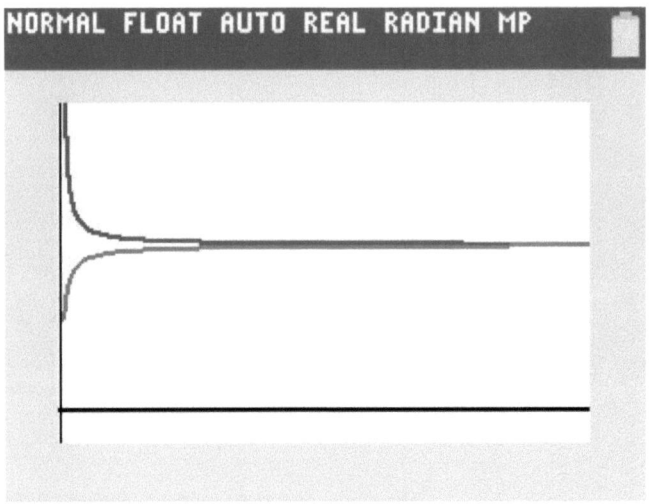

Graph 8 depicting $y = \left(1 + \frac{1}{x}\right)^x$ and $y = \left(1 + \frac{1}{x}\right)^{x+1}$ for a domain of x=0 to x=500

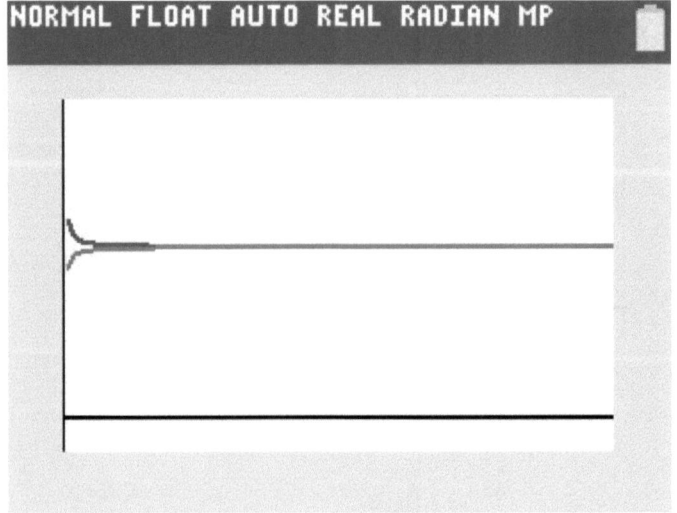

From the above graphs, it is evident that when the lower estimate is increased and when the upper estimate is decreased, their ratio will tend to 1. This means that these 2 functions will have a common limit when x is very large, that is at infinity. This limit has an approximate value of 2.7184178 when x is taken as 9999.

Graph 9: Value of e as the common limit of the two estimates at x=9999

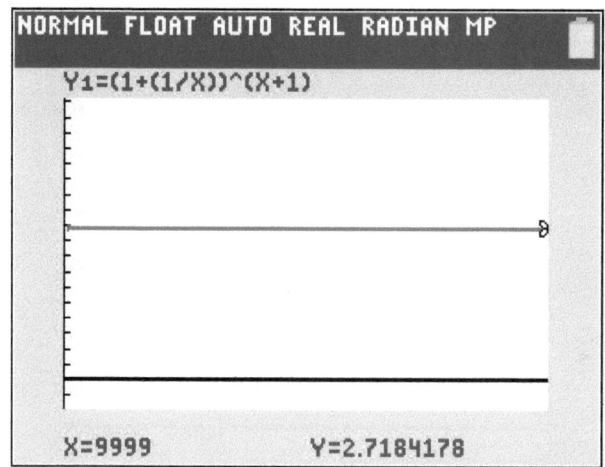

However, if I increase the domain to 100000, then the value of the limit comes out to be 2.7182954.

Graph 10: Value of e as the common limit of the two estimates at x=99999

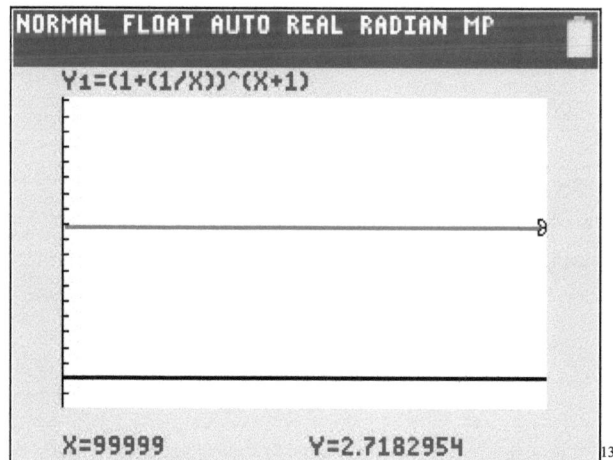

---

[13] Window for Graph10 in appendix.

If I further increase x to 1000000, the value becomes more precise and accurate. This yields a value of the limit as 2.7182832.

Graph 11: Value of e as the common limit of the two estimates at x=999999

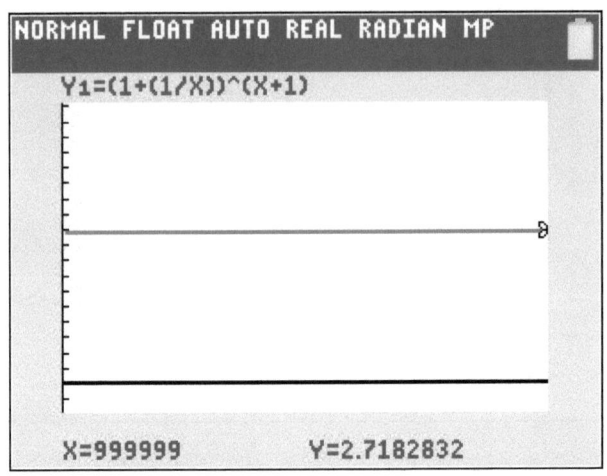

Here we see that as the value of x increases, or in other terms when the domain increases, the value of e comes more and more closer to true value that is 2.71828, when given up to 6 significant digits.

Thus, up to 6 significant digits:

$$\lim_{x \to \infty}\left(1+\frac{1}{x}\right)^x = \lim_{x \to \infty}\left(1+\frac{1}{x}\right)^{x+1} = 2.71828\ldots$$

This limit will have unending number of digits in the decimal part, just like $\sqrt{2}$ or π. Since it is not possible to write out the exact values of this number, it is designated by e in honor of the mathematician Leonhard Euler.[14]

---

[14] Morris Kline. "Calculus." *Calculus: An Intuitive and Physical Approach (Second Edition)*. Google Books, Web.
<https://books.google.co.in/books?id=tZy8AQAAQBAJ&pg=PA339&lpg=PA339&dq=why%2Bchoose%2Be%2Bas%2Bthe%2Bbase&source=bl&ots=uVPU2AsEqg&sig=nY0ZDKGrff4XewPx6AiDIGSHrxo&hl=en&sa=X&ved=0ahUKEwiKwN_gvd3KAhVBUo4KHY32BVUQ6AEIUTAJ#v=onepage&q=why%20choose%20e%20as%20the%20base&f=false>.

Conclusion

This means that arbitrary base a will now be termed as e and will have a value of approximately 2.718281828, which is correct up to 10 significant figures.
If we take this common limit or this value for the base a, for the exponential functions, an angle of 45° will be enclosed with the x-axis and the tangent at the coordinate (0,1).

<u>Graph 12 of e^x and $y = x + 1$</u>

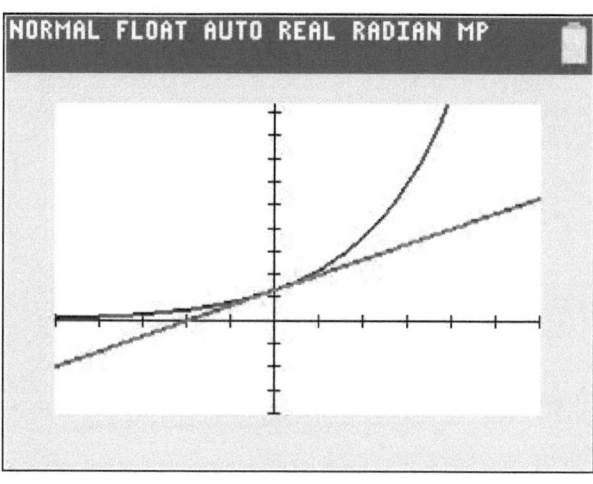

The graph above also stands as a testament only a base of e can make the slope of tangent of any exponential function at the coordinate (0,1) equal to 1. For this very reason the derivative of e^x is e^x itself. This is the most important property, and amalgamated with the fact that it is a transcendental number and has unending number of digits, make this 'natural'.

Therefore it is apt to say that the natural exponential function will have the base as *e*. With this established, a logarithm function is the inverse of the exponential function[15]. Therefore by inversing the natural exponential function we get a logarithm function with the base e, thus terming it as a natural logarithm function. A logarithm function can be defined as any increasing continuous function *f(x)*, defined on positive real numbers x, whose rate of growth

---

[15] Oberle. "Key Concepts." *Lrw.usd444*. Web. 2015.
<http://lrw.usd444.com/boberle/Mr._Oberle/Algebra_2_files/ch%208%20key%20concepts%20answers%20pdf.pdf>.

is inversely proportional to x. [16] This means that $\frac{df}{dx} \propto \frac{1}{x}$ where f is a logarithm function. Replacing the proportionality with a positive proportionality constant. For the logarithm to be a natural the constant must be 1. If it were the common logarithm the constant would have been something like 0.4342944819…this might have not been a very natural choice. The derivative of this natural logarithm function being $\frac{1}{x}$ can be called as a fundamental property of the natural logarithm. This natural function leads of to approximation formulae, that can't be possible for logarithm functions with other bases than e, like

$$\ln(1+x) = x - \frac{x^2}{2} + \frac{x^3}{3} + \cdots + \frac{(-1)^{n-1}x^n}{n} + R_n(x:0) \text{ [17]}$$

Such formulae can only be derived for natural logarithm functions as they give comparatively cleaner results.

The End.

---

[16] David Joyce. "Why Is the Logarithm with Base E Called a Natural Logarithm?" *Quora*. 23 Sept. 2014. Web. <https://www.quora.com/Why-is-the-logarithm-with-base-e-called-a-natural-logarithm>.

[17] Quinn, Catherine, C. J. Sangwin, R. C. Haese, and Michael Haese. "Taylor's Theorem." *Mathematics for the International Student: Mathematics HL (option): Calculus, HL Topic 9, FM Topic 5, for Use with IB Diploma Programme*. N.p.: Haese, 2013. 97. Print.

Bibliography:

1. Quinn, Catherine, C. J. Sangwin, R. C. Haese, and Michael Haese. "Taylor's Theorem." Mathematics for the International Student: Mathematics HL (option): Calculus, HL Topic 9, FM Topic 5, for Use with IB Diploma Programme. N.p.: Haese, 2013. 97. Print.

2. Oberle. "Key Concepts." Lrw.usd444. Web. 2015. <http://lrw.usd444.com/boberle/Mr._Oberle/Algebra_2_files/ch%208%20key%20concepts%20answers%20pdf.pdf>.

3. David Joyce. "Why Is the Logarithm with Base E Called a Natural Logarithm?" Quora. N.p., 23 Sept. 2014. Web. <https://www.quora.com/Why-is-the-logarithm-with-base-e-called-a-natural-logarithm>.

4. Morris Kline. "Calculus." Calculus: An Intuitive and Physical Approach (Second Edition). Google Books, n.d. Web. <https://books.google.co.in/books?id=tZy8AQAAQBAJ&pg=PA339&lpg=PA339&dq=why%2Bchoose%2Be%2Bas%2Bthe%2Bbase&source=bl&ots=uVPU2AsEqg&sig=nY0ZDKGrff4XewPx6AiDIGSHrxo&hl=en&sa=X&ved=0ahUKEwiKwN_gvd3KAhVBUo4KHY32BVUQ6AEIUTAJ#v=onepage&q=why%20choose%20e%20as%20the%20base&f=false>.

5. Quinn, Catherine, C. J. Sangwin, R. C. Haese, and Michael Haese. "Monotone Sequences." Mathematics for the International Student: Mathematics HL (option): Calculus, HL Topic 9, FM Topic 5, for Use with IB Diploma Programme. N.p.: Haese, 2013. 64. Print.

6. Quinn, Catherine, C. J. Sangwin, R. C. Haese, and Michael Haese. "Monotone Sequences." Mathematics for the International Student: Mathematics HL (option): Calculus, HL Topic 9, FM Topic 5, for Use with IB Diploma Programme. N.p.: Haese, 2013. 64. Print.

7. Martin, David. "Differentiation from First Principles." Mathematics for the International Student: Mathematics HL (Core). 3rd ed. Adelaide: Haese Mathematics, 2012. 523. Print.

8. Zidar, Owen. "A Primer on Derivatives and Maximization Problems." A Primer on Derivatives and Maximization Problems 1 (n.d.): n. pag.Faculty.chicagobooth. Chicagobooth, 2015. Web. 2015. <http://faculty.chicagobooth.edu/owen.zidar/teaching/Fall%202015/Week0/Week0_Primer_On_Derivatives_and_Maximization.pdf>.

9. Kahen. "Show That D/dx (a^x) = $A^x \ln A$." Calculus. Stack Exchange Inc, May 2013. Web. 2015. <http://math.stackexchange.com/questions/398826/show-that-d-dx-ax-ax-ln-a>.

10. "Derivative of Exponential Functions." Derivative of Exponential Functions. Geogebra, n.d. Web. 2015.

<http://webspace.ship.edu/msrenault/GeoGebraCalculus/derivative_exponential_functions.html>.

11. O' Conner, J J, and E F Robertson. "The Number E." MacTutor History of Mathematics, Sept. 2001. Web. 2015.
12. Gelman, Andrew, and Jennifer Hill. Data Analysis Using Regression and Multilevel/Hierarchical Models. 1st ed. New York: Cambridge UP, 2007. 60-61. Print.
13. Cook, John D. "Another Reason Natural Logarithms Are Natural." Data Analysis. John D. Cook, 5 Feb. 2015. Web. 2015. <http://www.johndcook.com/blog/2015/02/05/natural-logarithms-are-natural/>.
14. Pickover, Cliff. "The 15 Most Famous Transcendental Numbers – Cliff Pickover".Sprott.physics.wisc.edu. Web. 2015.

Appendix

Inputs for graph 1

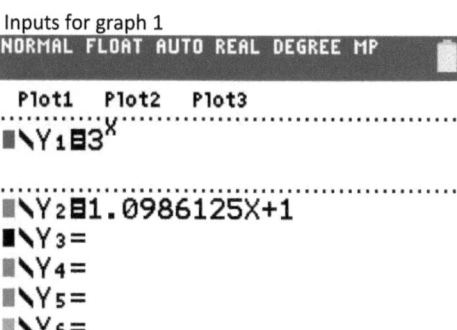

I first took the derivative of the curve 3^x, and then found out the equation of the tangent at point (0,1).

Inputs for graph 5

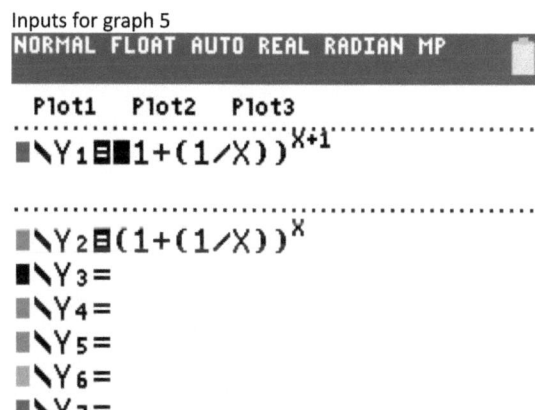

Window for Graph 10

```
NORMAL FLOAT AUTO REAL RADIAN MP

WINDOW
 Xmin=-.5
 Xmax=100000
 Xscl=25
 Ymin=-.5
 Ymax=5
 Yscl=.275
 Xres=1
 ΔX=378.78977272727
 TraceStep=757.57954545454
```

YOUR KNOWLEDGE HAS VALUE

- We will publish your bachelor's and master's thesis, essays and papers

- Your own eBook and book - sold worldwide in all relevant shops

- Earn money with each sale

Upload your text at www.GRIN.com
and publish for free